Red and Green In

SPRING

Bonnie Carole

Educational Media

rourkeeducationalmedia.com

Scan for Related Titles
and Teacher Resources

Teaching Focus:

Have students find capital letters and punctuation in a sentence. Ask students to explain the purpose for using them in a sentence.

Before Reading:

Building Academic Vocabulary and Background Knowledge

Before reading a book, it is important to set the stage for your child or student by using pre-reading strategies. This will help them develop their vocabulary, increase their reading comprehension, and make connections across the curriculum.

1. Read the title and look at the cover. *Let's make predictions about what this book will be about.*
2. Take a picture walk by talking about the pictures/photographs in the book. Implant the vocabulary as you take the picture walk. Be sure to talk about the text features such as headings, Table of Contents, glossary, bolded words, captions, charts/diagrams, or Index.
3. Have students read the first page of text with you then have students read the remaining text.
4. Strategy Talk – use to assist students while reading.
   - Get your mouth ready
   - Look at the picture
   - Think…does it make sense
   - Think…does it look right
   - Think…does it sound right
   - Chunk it – by looking for a part you know
5. Read it again.
6. After reading the book complete the activities below.

High Frequency Words

Flip through the book and locate how many times the high frequency words were used.

and
are
green
is
it's
red
spring
the
they're
what

After Reading:

Comprehension and Extension Activity

After reading the book, work on the following questions with your child or students in order to check their level of reading comprehension and content mastery.

1. *What kinds of berries are red?* (Summarize)
2. *Why are raincoats worn in the spring?* (Asking questions)
3. *What is your favorite activity to do in the spring?* (Text to self connection)
4. *Why is spring a fresh start for nature?* (Inferring)

Extension Activity

Watch a plant grow! Spring brings sun and rain which makes plants grow. You can watch a plant grow from your window. You need a plastic sandwich bag, a paper towel, a bean seed, and masking tape. With the help of an adult, get the paper towel wet and fold it twice to fit inside the plastic sandwich bag. Next place one bean seed in between the plastic bag and wet paper towel. Seal the bag and tape it to a window that gets a lot of sun. Each day use a notebook and record what you see. How does the warm sun and water change the seed? Draw and label your seed each day.

When you think of spring, do you think **red** and green?

Spring brings something new,
the world comes alive for me
and you!

Red or green? What color is the grass?

It's green!

Red or green? What color is the ladybug?

It's **red**!

Red or green? What color
are the berries?

They're **red**!

Red or green? What color is the kite?

It's green!

Red or green? What color are the stitches on the baseball?

They're **red**!

Red or green? What color is the swing at the park?

It's green!

Red or green? What color are the flowers?

They're **red**!

Red or green? What color is the raincoat?

It's **red**!

Red or green? What color is the turtle?

It's green!

Spring is a fresh start for everything!

Index

green 3, 5, 6, 7, 9, 11, 12, 13, 15, 16, 17, 19, 21, 22

park 15

raincoat 19

red 3, 5, 7, 8, 9, 10, 11, 13, 14, 15, 17, 18, 19, 20, 21

spring 3, 4, 23

Websites

http://pbskids.org/catinthehat/games/chasing-rainbows.html

http://www.scholastic.com/clifford/play/seasonalstickers/stickers-main.swf

http://pbskids.org/sid/weatherwheel.html

Meet The Author!
www.meetREMauthors.com

About the Author

Bonnie Carole lives in Illinois with her family and three dogs. They love going to the pond and seeing all the baby ducks and geese in the spring.

www.rourkeeducationalmedia.com

PHOTO CREDITS: Cover illustrated type © Melica, cover photo © Zoia Kostina; title page and page 8 ladybug © irin-k; page 3 © Destinyweddingstudio; page 4-5 © Africa Studio; page 6 © photolinc; page 7 © yanikap; page 9 © Tatiana Volgutova; page 10 © Volosina; page 11 © Rawpixel; page 12 © francesco de marco and Rawpixel; page 13 © momento; page 14 © Pincarel; page 15 and 16 © Photographee.eu; page 17 and 18 © Zoia Kostina; page 19 © BlueOrange Studio; page 20 © Roman Sigaev; page 21 © Vitaly Titov & Maria Sidelnikova; page 22 © Archiwiz; page 23 © Cheryl E. Davis

Edited by: Luana Mitten

Cover design and Interior design: by Nicola Stratford
www.nicolastratford.com

Library of Congress PCN Data

Red and Green in Spring / Bonnie Carole
(Concepts)
ISBN 978-1-63430-048-3 (hard cover)
ISBN 978-1-63430-078-0 (soft cover)
ISBN 978-1-63430-106-0 (e-Book)
Library of Congress Control Number: 2014955191

Rourke Educational Media
Printed in the United States of America, North Mankato, Minnesota

Also Available as: